MATH
and
ME

Math at School

by Joanne Mattern

LOOK!
BOOKS™

Red Chair Press   Egremont, Massachusetts

Look! Books are produced and published by Red Chair Press:

Red Chair Press LLC   PO Box 333   South Egremont, MA 01258-0333

FREE Educator Guides at www.redchairpress.com/free-resources

Publisher's Cataloging-In-Publication Data
Names: Mattern, Joanne, 1963- author.
Title: Math at school / by Joanne Mattern.

Description: Egremont, Massachusetts : Red Chair Press, [2022] | Series:
   LOOK! books : Math and Me | Interest age level: 005-008. | Includes
   index and suggested resources for further reading. | Summary: "There
   are lots of ways to use math at school. Readers will count students and
   objects, learn about time, and use addition and subtraction as they go
   through a typical day at school"--Provided by publisher.

Identifiers: ISBN 9781643711294 (hardcover) | ISBN 9781643711355
   (softcover) | ISBN 9781643711416 (ePDF) | ISBN 9781643711478 (ePub 3
   S&L) | ISBN 9781643711539 (ePub 3 TR) | ISBN 9781643711591 (Kindle)

Subjects: LCSH: Mathematics--Juvenile literature. | School day--
   Mathematics--Juvenile literature. | CYAC: Mathematics. | Schools.

Classification: LCC QA40.5 .M382 2022 (print) | LCC QA40.5 (ebook) | DDC
   510--dc23

Library of Congress Control Number: 2021945362

Photo credits: Cover, p. 3, 5, 7, 8, 10, 11, 15, 17–22: iStock; p. 1, 6, 13, 23, 24:
Shutterstock

Printed in United States of America
0422 1P CGF22

Table of Contents

Let's Go to School 4

Getting to School 6

Time for Class 10

How Many Students?. 12

Working in Groups. 14

Science Class 16

Lunch Time!. 18

Going Home 20

Words to Know. 23

Learn More at the Library 23

Index. .24

About the Author. 24

Let's Go to School

You use many **skills** at school. You learn math during class. But there are lots of other ways you use math when you are at school. Let's take a look at some ways math is part of the school day. You can have fun there too!

Getting to School

Students get to school in many ways. Some students take the bus. Some students ride their bikes. Some students walk to school. Some students get a ride from their family.

Five students in this class take the bus to school. Two students walk. One student gets a ride from a parent. Four students ride their bikes. How many students are in the class?

9

Time for Class

These students got to school at 8:15. Class starts at 8:30. The students talk and play before class starts. How much time do they have?

8:15 is 15 minutes before 8:30. The students have 15 minutes to play.

How Many Students?

Here is the teacher. He needs to know how many students are in class. He calls each student by name. There are 15 students in the class. Three students are absent. How many students are in class today?

You can write the question like this:

15 – 3 = 12

There are 12 students in class today.

MATH FACT!

Working in Groups

The first class is reading. The teacher divides the class into two **equal** groups. How many children are in each group? Let's count.

There are six children in each group.

Science Class

It is time for science class. The students pour water into measuring cups. One cup holds 12 **ounces**. The other cup holds 8 ounces. Which cup holds more?

Twelve is greater than eight. You can write this as 12 > 8. The 12-ounce cup holds more.

MATH FACT!

Lunch Time!

The students work hard all morning. Now it is 12:00. It is time for lunch. Eight students buy lunch. If there are 12 students in class, how many bring their lunch from home?

MATH FACT!

Subtract 8 from 12 to get the answer.

12 − 8 = 4

Four students brought lunch from home.

Going Home

The students have worked hard all day. Now it is 2:45. It is time to go home. There is a long line for the bus! How many children are in line? Let's count.

1, 2, 3, 4, 5, 6, 7, 8.

The students learned a lot today. They used math in many different ways. See you back at school tomorrow!

Words to Know

equal: the same

ounces: units of weight-measurement

skills: abilities to do something

students: people who go to school

Learn More at the Library

Check out these books to learn more.

Levit, Joe. *Let's Explore Math (Bumba).* Lerner Publications, 2019.

Overdeck, Laura. *Bedtime Math Series.* Feiwel and Friends, 2013.

Nielsen, Aubrie. *At School (Mathematics Readers).* Teacher Created Materials, 2020.

Index

bikes 6, 9

bus 6, 9

lunch 18

reading class 14

science class. 16

teacher 12, 14

About the Author

Joanne Mattern is the author of many books for children. She loves writing about sports, animals, and interesting people. Mattern lives in New York State with her family.

24